Kingfell Guide

KF913

Second Edition: 2013

Fire risk assessments for complex buildings

Fire risk assessments for complex buildings

ISBN: 1489541144
ISBN-13: 978-1489541147

CONTENTS

ACKNOWLEDGMENTS

I would like to thank my colleagues at Kingfell, past and present, for the some of the ideas used to write this Guide. I would also like to thank colleagues at London Underground, Network Rail (UK), the Houses of Parliament, London Fire Brigade and those I meet at IFPO (Institute of Fire Prevention Officers) meetings in Southwark Cathedral, London. It was their inspiration and support that put me on the road to preparing a document covering fire risk assessments specifically for complex buildings.

Paul Bryant

PREAMBLE

The fire risk assessment is increasingly becoming a fundamental requirement and process in many parts of the world. The primary aim of the assessment is to ensure fire safety and protection provisions are fit for purpose and that any fire hazards are suitably controlled and/or protected against.

In 2005, updated United Kingdom fire safety legislation introduced the fire risk assessment as a primary legal requirement for most building types other than domestic dwellings. This was welcomed as a departure from inflexible and possibly inappropriate prescription. However, since the introduction of the legislation, there have been concerns raised as to the consistency and quality of the assessments.

As the term suggests, a fire risk assessment is a direct assessment of the fire risks pertaining to a building or other structure. These risks are then identified and, typically, a list is drawn up of actions and recommendations to mitigate the risks by fire prevention and protection techniques.

From a sample of fire risk assessment formats, it could be said that some are akin to fire *compliance* assessments. They compare the current fire safety provisions of the building with national fire safety and building requirements. Any areas that vary from the requirements are identified and recommendations are made where the provisions full short of the requirements. It is quite possible that this form of assessment never actually looks at the *risk* of fire, and its implications.

Another issue that concerns many is the competency to undertake fire risk assessments. Even though schemes are being developed to *assess* the assessors, the author of this Guide believes that competency to undertake mainstream fire risk assessments is one thing, but assessing buildings and infrastructures that could be described as complex, is another.

Then there is the scope of the assessment. The vast majority of fire risk assessments consider primarily the life safety risk. This is most commonly due to the need to meet with national fire safety legislation, which tends to focus on life safety. With complex buildings, it is possible that issues such as asset protection, business continuity and environmental protection are also relevant. Would not this be an ideal opportunity for such objectives to also be investigated by the risk assessment process?

Because of these reasons, the author, Paul Bryant, decided to look more deeply into the subject using his knowledge and experience of the preparation of fire strategies, mostly for complex buildings.

This Guide incorporates both the results of his findings and his views on the subject. This edition has been completely reformatted from the first edition, for publication both as a printed book and as an ebook. It has also been changed from a Code to a Guide to reflect its advisory status. Parts of the text have been amended and additional information has been included. The assessment table in Annex C has been updated

This Guide has been developed by the author for Kingfell, which retains its ownership and copyright. It does not purport to include all necessary provisions of a contract. Users are responsible for its correct application. Furthermore, Kingfell accepts no responsibility or liability where this document was partially or wholly used for the undertaking of fire risk assessments.

This edition is KF913 – Second edition: 2013. This Guide uses UK English spelling.

INTRODUCTION

Fire risk assessments are becoming a fundamental part of fire safety practice around the world. They are the basis of current UK fire safety legislation; the UK's Regulatory Reform (Fire Safety) Order 2005.

The purpose of the risk assessment is to ensure that fire precautions are applied commensurate with the risks found within the building. This is a departure from the prescriptive approach typically adopted by many countries including that of former UK legislation, when a set of measures were prescribed by the enforcer (the fire authority) as detailed in a fire certificate. The purpose of the current UK legislation is to ensure fire safety management and protection systems are directly applicable to the building and its occupancy and allows greater degrees of flexibility in specifying and applying such management and protection systems. This idea is transferable to many parts of the world.

In general, there is little absolute guidance as to how a risk assessment should be undertaken, and what it should include, although the basic aspects of the assessment, notably the basic fire prevention measures, the means of escape and means of raising an alarm, are normally included.

This is satisfactory in many instances where such provisions are simple and straightforward. However, where the building is more complex, perhaps containing multiple floors, incorporating operational processes or is used by special groups such as the general public, then the fire safety and fire protection provisions are also likely to be more complex.

The purpose of this document is to give additional guidance for those undertaking fire risk assessments when faced with a set of fire precautions, management controls and fire protection systems that cannot be easily and readily verified for compliance, operability and effectiveness.

SCOPE

This document provides guidance for fire risk assessors for the assessment of fire risks in complex buildings. In particular, it contains considerations to assess the suitability and effectiveness of fire safety and protection measures for buildings or other sites where provisions may not be straightforward to assess and / or where a more thorough analysis is required. This will mostly be the case for those buildings or sites that can be described as complex based on size, layout complexity, internal processes and occupancy profiles.

It includes considerations for fire risk assessments for buildings that have used a performance based/fire safety engineered approach when determining fire safety and protection.

It will be assumed that the risk assessor already has a fundamental understanding of risk and hazard assessment. This document has been prepared to provide additional guidance.

This document is written to apply to most types of organisation and all types of complex building and infrastructure and should be used as deemed appropriate to the circumstances of the organisation.

TERMS AND DEFINITIONS

For the purposes of this Guide, the following terms and definitions apply.

ASET (Available safe escape time) - Calculated time available between the ignition of a fire and the time at which the conditions for escape become untenable to the occupants.

Compartmentation - Sub-division of a building by fire-resisting walls and/or floors.

Competent Person - A person with the necessary skills, training and experience to undertake fire risk assessments.

Complex Building - The premises that is the subject of the fire risk assessment. Note that this could also refer to a group of buildings or those with complex layouts and uses such as mines and onshore or off-shore oil and gas processing plants.

Concealed space or cavity - Space enclosed by elements of a building (including a suspended ceiling) or contained within an element, but not a room, cupboard, circulation space, protected shaft or space within a flue, chute, duct, pipe or conduit.

Emergency lighting - Lighting provided for use when the supply to the normal lighting fails.

Escape route - Route forming part of the means of escape from any point in a building to a final exit.

Final exit - Termination of an escape route from a building giving direct access to a street, passageway, walkway or open space, and sited to enable the rapid dispersal of persons from the vicinity of a building so that they are no longer in danger from fire and/or smoke.

Fire damper - Mobile closure within a duct, which is operated automatically or manually and is designed to prevent the passage of fire and which, together with its frame, is capable of meeting for a stated period of time the fire resistance criterion for integrity.

Fire door - Door or shutter provided for the passage of persons, air or objects which, together with its frame and furniture as installed in a building, is intended (when closed) to resist the passage of fire and/or gaseous products of combustion, and is capable of meeting specified performance criteria to those ends.

Fire resistance - Ability of a component or construction of a building to meet certain integrity and insulation requirements, when exposed to a fire on one side, for a stated period of time.

Managed evacuation - Process where evacuation is managed by trained persons using on-site assistance, coded warnings etc.

Management of fire safety - Tasks carried out by a defined individual or individuals with appropriate powers and resources to ensure that the fire safety prevention and protection systems, procedures and processes within the building are working properly at all times.

Means of escape - Structural means whereby a safe route in the event of fire is provided for persons to travel from any point in a building to a place of safety.

Phased evacuation - Process where a limited number of floors (or zoned horizontal areas) are evacuated at the same time. For vertical evacuation, this is usually the floor containing the fire and the floor above. The remaining floors (or horizontal areas) are evacuated at later stages if required.

Place of safety - Place in which persons are in no immediate danger from the effects of a fire.

Place of relative safety - Place in which persons are in no immediate danger from the effects of a fire for a specific period and from there can be moved to a place of safety.

Pre-movement time - Interval between the time at which a warning of fire is given and the time at which the first move is made towards an exit.

Responsible person - The person ultimately deemed responsible for the fire safety of the building by virtue of legislation and/or other reasons as required by stakeholders.

RSET (Required safe escape time) - The time required, measured from the ignition of a fire, for all occupants of a building to reach a place of safety. This includes the time of the alarm being raised, pre-movement time and total escape time.

Smoke control - Technique used to control the movement of

smoky gases within a building in order to protect the structure, the contents, the means of escape, or to assist fire-fighting operations.

Smoke damper - Mechanical device which, when closed, prevents smoke passing through an aperture within a duct or structure.

Stakeholder - Person or organisation having a vested interest in the fire safety and protection of the building being assessed.

Travel distance - Actual distance a person needs to travel from any point within a building to the nearest exit, having regard to the layout of walls, partitions and fittings.

1   GENERAL CONSIDERATIONS

Prior to undertaking the assessment, it is important that those involved fully appreciate and understand the purpose of the risk assessment. Although the primary benefit is normally to achieve and ensure compliance with fire safety legislation, there are ethical and other factors to be considered that are especially relevant for complex buildings. These include:

- To consider the requirements and needs of other stakeholders such as the insurer, local authorities, enforcers and special groups with a vested interest in the building, its occupancy and its processes.

- To prompt a full and thorough consideration of the fire safety requirements of the premises in question and of its occupants as part of a wider risk management strategy.

- To widen the consideration of the impact of fire precautions to broader objectives encompassing life safety, property protection, business continuity/protection and the environment.

- To review fire system design criteria prior to the specification of new fire protection systems.

- To incorporate any requirements of the fire and rescue and civil defence authorities should they have requirement to obtain specific information relating to the building being assessed.

What is a Complex Building?

A complex building is difficult to define absolutely. For the purposes of this document a complex building incorporates one or more of the following features:

- It is large in size and /or incorporates many floors.

- It houses complex processes or hazards that may require specialist protection systems.

- It incorporates occupancy profiles that may be large, varied or will have minimal knowledge of their environment.

- The fire strategy and subsequent designs are based upon a fire safety engineered solution and not on a recognised prescriptive approach.

Competency to undertake the risk assessment

It is vital that those entrusted to undertake a fire risk assessment for complex buildings have the necessary levels of competence to undertake the task professionally and thoroughly. The level of competency required will be commensurate with the expected complexity of the building to be assessed, but the person, or team should have the following credentials:

- To have a good understanding of fire related aspects of building control and function.

- To have an appropriate knowledge of national fire legislation and the requirements of other enforcing bodies, as well as an understanding of stakeholders such as insurers.

- To be appropriately trained and / or experienced in fire safety and fire protection issues.

- To be knowledgeable of relevant national and local codes and have had past experience of their application.

One way of validating competency is to ensure that those undertaking the fire risk assessment are approved or accredited by a relevant third party professional organisation. Note that even accredited or approved persons or organisations may not have the skill set to evaluate some of the more detailed aspects of complex building fire precautions and systems. This should be ascertained from the approving organisation or from the fire risk assessor prior to their engagement.

Understanding the objectives for the assessment

Typically the primary purpose of the fire risk assessment is to ensure that fire provisions have been, or should be, applied commensurate with the life safety risk of the building.

Some national legislation acknowledges purposes other than life safety but in the case of complex buildings, the assessment is an ideal opportunity to look at the wider objectives of life safety, property protection, business continuity and protection of the environment against the impact of a fire.

For each of these categories, there will be a number of issues which may include:

Life safety: On top of the needs of fire safety legislation, there should be an ethical need to ensure the safety of all occupants. A good example is where the general public occupy the building environment. There is an obligation to ensure maximum safety and the consequences of exposing the public to even a heightened risk of fire will be judged to be unacceptable.

The safety of firefighters is another example of where the assessor can expand the scope to cover an aspect required outside of regulatory compliance. In some countries, fire and civil authorities are obliged to undertake risk assessments of certain building types and profiles as part of their remit. It may be possible that the requirements of this fire risk assessment are combined with their general risk assessment.

Property Protection: Insurers may impose specific requirements to protect the building from a fire. These will normally be aimed at restricting and containing fire growth and where possible suppressing and/or extinguishing the fire to minimize direct and indirect damage (such as from smoke). Consequently, there may be a greater reliance on appropriate fire compartmentation and fire suppression systems such as automatic sprinkler systems.

Business Continuity and Protection: Many businesses collapse after suffering from a fire. In many cases this is not immediate but due to the loss of short term operations which can damage both client and supplier confidence. This can lead to a gradual demise in trading. Business continuity usually points to resilience built into operations via redundancy, etc. but certain processes, equipment and systems may be vital and hard to duplicate in the short term. In such cases, additional or specialised fire protection may be required.

Environmental Protection: It is increasingly recognised that a fire can have a huge and long lasting detrimental effect to the local and wider environment. This can occur in many ways, for example:

- Products of combustion from the fire causing damage to the local environment.

- Products of combustion incorporating materials that may cause health risks to those nearby.

- Power and water supplies becoming damaged or contaminated and impacting on the local community.

- Pollution of local water supplies and rivers due to the released products of combustion or due to the firefighting water effluent.

Given the heightened awareness of litigation, it may be advisable that environmental issues are considered as part of the scope of the assessment.

These examples may not form part of a general fire risk assessment but when included as part of a detailed assessment can provide a more meaningful result and highlight issues that may not have been fully appreciated.

2 PROCEDURE FOR UNDERTAKING THE FIRE RISK ASSESSMENT

Various risk assessment processes can been used and many national, regional and sector based formats for the recording of the assessment may be applicable. However, for complex buildings, issues such as the initial assessment of risk (including the identification of hazards) and the determination of appropriate fire safety provisions, should have been undertaken at some point prior to the assessment, possibly at the building design stage and then followed up by inspections from building control authorities or other bodies as appropriate. It is also probable that complex buildings would have been issued with approval or certification for the fire safety measures taken. This was the case in the UK prior to the current legislation, where fire certificates were issued.

It should be expected that the risk and hazard assessment of the building will not normally be an assessment from first principles but a follow on from prior analysis contained in fire strategies, agreements, designs etc.

Supporting Documentation prior to the assessment

As described above, complex buildings are likely to have undergone a full and thorough assessment with respect to fire safety and protection. Consequently, there may be a number of documents relevant to the assessment that should be reviewed prior to the assessment. These may include:

- The fire strategy document together with updates as appropriate.

- The requirements of any prior fire certificates or approvals (where applicable).

- The fire safety manual.

- The requirements of any other stakeholders such as the insurer.

- Fire protection system designs and configuration information (such as cause and effect programming).

- Drawings and layouts showing fire compartment lines and fire protection system locations.

- Minutes and reports of corporate fire safety reviews and meetings.

- Supporting information for performance based designs such as fire and evacuation models, etc.

Risk reduction and "ALARP"

The risk assessment should take into account risks described as "ALARP" (**A**s **L**ow **A**s **R**easonably **P**racticable). Risks described as such may not require special provisions or consideration if all parties agree that this level has been reached for specific aspects of the building. The assessor should be aware of this and the reasoning behind such instances.

The scope and format of the fire risk assessment

The fire risk assessment may require consideration of much more than that typically associated for the purpose of complying with legislation. It is therefore recommended that the assessor or the assessment team agree the scope of the assessment prior to the undertaking of the assessment itself. Any special conditions, assumptions or acceptance criteria should also be discussed.

With regard to the format of the inspection, a fire risk assessment tick list or one that uses a scoring mechanism is unlikely to be sufficient on its own. By definition, a complex building will contain aspects that will not have a "yes or no" answer and there will be a need to define and describe what has been found and why and how it complies or does not comply with that which is deemed necessary.

3   STAGES IN THE ASSESSMENT PROCESS

Typically, a systematic fire risk assessment review should be conducted to establish the fire related hazards within the building and their potential consequences. For complex buildings, the following staged process could be followed, although this may be amended subject to the requirements of stakeholders. See Annex B for the flow chart.

TASK 1. Pre-assessment meeting

It would be helpful, prior to undertaking any work, to meet with the client and/or responsible person of the premises. At this meeting, the key requirements can be ascertained. This may include their actual objectives and scope for the assessment. Note that questioning may reveal a need to look wider than the basic requirement to meet legislation as described earlier. This can also be a good time to gather relevant documentation.

Task 2. Documentation gathering and inspection

Following the meeting, all relevant documentation should be gathered and assessed to determine the risks and hazards as perceived and how the risks were, and are, mitigated. (Note that more detailed considerations are given later in this Guide). The following list incorporates some of the information that may be deemed to be useful and relevant:

- Fire safety management and training details.

- The provisions for first aid and professional firefighting.

- Where appropriate, the key performance requirements, where a fire safety engineered solution was used. This may include ASET and RSET calculations and the output of fire and evacuation modelling and how these may impact on those calculations.

- Controls on escape routes including dimensioning, distances allowed, horizontal and vertical parameters and special requirements for refuges, relative places of safety etc.

- Passive fire protection requirements and how these support the prescriptive or performance based designs.

- History of previous fires and evacuations.

- Interrelationships between the fire strategy and security and other protection strategies.

- Information covering the active fire protection systems, including details of how these support the prescriptive or performance based designs. Design categories and criteria should also be assessed as being fit for purpose.

- Where appropriate, property and asset protection methodologies as specified by stakeholders such as the insurer.

- Where appropriate, systems or arrangements to ensure business continuity in the event of fire.

- Where appropriate, systems or arrangements to protect the local and wider environment in the event of fire, or from the effects of fighting the fire.

Task 3. Gap analysis

One important outcome of the documentation inspection is to evaluate if there are gaps between the hazards and risks as now identified and the original provisions applied. This gap could be due to modifications to the building design or layout, changes to the occupancy profile to that originally envisaged or changes in the use of part, or all, of the building. The gaps analysis will help the assessor to determine if the provisions, as seen, are still suitable, or that there may need to be modifications or enhancements to that provided. The gap analysis is also an ideal opportunity to re-evaluate the fire strategy for the building.

Task 4. Scope and timetable agreement

From the earlier analysis, a scope can be drawn up and agreed with the client or nominated responsible person and any other persons affected by the assessment. It would be helpful to draw up a timetable of activity to ensure that persons deemed necessary to support the assessment are available.

Task 5. The on-site assessment

The following should be considered when assessing on the site:

- Suitable time should be allowed to completely cover the premises, including the walking of all escape routes from worst case internal points through to final places of safety.

- It is recommended that all rooms are inspected and those containing processes are properly evaluated.

- The assessor should evaluate the true nature of the occupancy profile(s) and whether the assumptions found by earlier research stand up to reality.

- It may be helpful if the assessor ask relevant questions directly to a sample of building occupants. Prior agreement to this should be sought from the client and / or responsible person. This questioning can help test the robustness of the fire precautions, particularly those concerned with management of fire safety.

- Fire protection systems should be evaluated. This should include noting obvious deficiencies and also an assessment of the control and indicating equipment and the associated maintenance and operational log books.

- Firefighting arrangements, from portable fire extinguishers through to fire risers, should be evaluated and any evidence of poor maintenance or incorrect locations, identified.

- Arrangements for emergency services, from access through to on-site procedures, should be evaluated.

- All special hazards, observations or non-compliances should be noted and recorded, preferably with a photo record. It is recommended that a written description is used in all cases as a scoring and/or tick box system may not pick up the pertinent features when reviewed at a later date.

Task 6. Completion of the on-site assessment

Once the on-site assessment is completed, the client and / or responsible person, or their representative(s), should be notified. If the outcome is requested on the day of the assessment then it should be caveated that the assessment is not fully complete until the items found have been fully considered.

Note that post assessment tasks are described in Chapter 5.

4   DETAILED CONSIDERATIONS

Listed below are a number of questions that may be relevant during the assessment process, either at the information gathering stage, during the assessment, or part of the post assessment evaluation. Note that the list may not be exhaustive. Similarly, some of the considerations may not be relevant to the building in question. Detailed guidance may be found in national standards.

Fire safety management

Fire safety management is key to ensuring that fire safety procedures, processes and protection systems are likely to be appropriate to the risk and will be ready and effective in the event of fire. As well as the need to manage fire safety for legislative or other reasons, the corporate culture towards fire risk in general will assist in guiding the assessor towards potential issues. The following considerations will be helpful in determining this.

- Are there details of responsible persons for fire safety?

- Is there a clearly stated reporting structure for fire safety including procedures for review and authorisation?

- Are there specific policies for fire prevention and housekeeping including the control of processes and hot works? Is there any evidence that these are adhered to?

- Are fire drills undertaken? Are these effective and do they result in a review?

- Are there procedures for the control of maintenance of fire protection systems? Is there any evidence that these are adhered to?

- Is there any form of training for fire safety including warden training, use of portable fire extinguishers and the training of contractors for working on site?

Building characteristics

The characteristics of the building in terms of its location, construction, etc. should always form part of the assessment.

Although a detailed analysis of the building design is not usually necessary, the assessment should consider at least the following points:

- Are there any aspects with regard to the location of the building, or features of the building, that could impede evacuation in the event of a fire?

- Is there a clear strategy with regard to the arrangement of internal fire compartments and separations? Does the site assessment confirm this arrangement?

- Are openings within the building such as voids, that could allow routes for uncontrolled fire or smoke spread, controlled or protected by structural means or by system monitoring and containment?

- Are the fire properties of external and internal walls and linings, floors, ceilings and roof coverings stated and does the site assessment confirm these to be accurate?

- Can it be confirmed that nominated escape routes can be kept clear for the duration of the evacuation ?

- What forms of natural ventilation exist ? Are they suitable for the expected type and quantities of smoke envisaged? If not smoke extract systems may be required (see later).

- What is the potential for fire and smoke to spread through unknown or concealed routes? If so, how can these be controlled?

Building plant and engineering services

Building plant and engineering services are obvious candidates for a more detailed risk and hazard assessment. The following gives some guidance:

- How susceptible are key services to a fire? In this case, location and separation from other risk areas will need to be assessed.

- How could plant contribute to fire, smoke and toxic gas movement throughout the building (such as may be the case with ventilation)? Has this been catered for by passive or active fire protection?

- How easy will it be to access critical services in the event of a fire? As part of this assessment, the location of fire detection and alarm control and indicating equipment to control smoke extraction should be considered. In this case, the relative location and accessibility to means of escape or other safe area will need to be assessed.

Building processes

All buildings contain a set of processes. These may be tangible and obvious or more subtle and part of the environment of the building. Key processes may include IT systems, manufacturing lines, storage facilities and equipment, chemical procedures, etc. Similarly, the handling of people at airports and railway stations could be regarded as a building process. The following gives some guidance:

- How susceptible are key processes to a fire? In this case, location and separation from other risk areas will need to be assessed.

- How could each of the processes contribute to a fire or to smoke and toxic gas movement throughout the building ? Has this been catered for by passive or active fire protection?

Occupancy characteristics

The characteristics of each of the occupancy groups, within any building, will impact on the overall life safety fire risk. Consequently, it is recommended that each occupancy group is separately evaluated. Occupancy Groups may include:

- Employees who would normally have knowledge of the building and the evacuation arrangements.

- Contractors who may have received fire training in the building. However, additional consideration will need to be given to the locations where they may be during a fire incident. In some cases, they may have to dismount from access equipment or retract themselves from restricted environments prior to being able to evacuate from the building. This additional time required should be evaluated as part of the assessment.

- Visitors – occasional or regular

- The general public

- Those with special access needs

  Considerations for each group could include:

- Does the assessor believe that the above groups are adequately covered by the documentation assessed prior to the site inspection and /or as noted during the inspection?

- Are the majority of occupants knowledgeable of the building, its layout and possibly fire and other exit routes from the building?

- Is the ability range to evacuate in a fire largely known and covered by the provisions specified?.

- Is "pre movement time" understood and/or defined?

- Is there a likelihood that the density of occupants, such as visitors, may vary greatly within the building, with a greater concentration in parts? If so, are the escape provisions acceptable?

- Are there language and comprehension of alarm message issues and have these been taken into account?

Evacuation and Means of Escape

A fundamental requirement of a fire risk assessment for legislative reasons is to ensure that safe escape of occupants of the building during a fire. These are some of the main considerations:

- Is there an evacuation strategy? How is it documented?

- For performance based designs, are the ASET and RSET figures defined? Is the difference in values sufficient based on the findings of the assessment? Furthermore, Do the ASET and RSET figures allow for aspects such as pre-movement time and pre-evacuation procedures and coded warning systems?

- How have the occupancy numbers and profiles been established? Are they a true representation of what is seen during the assessment?

- What assumptions have been made about spread of persons throughout the building? Is there a high possibility of skewing of numbers into particular parts of the building?

- Are the horizontal and vertical means of escape clearly identified on drawings? Does the site assessment confirm the accuracy of these drawings?

- Are there any unusual aspects of the means of escape that demand additional attention (such as travel distance, dead end conditions, etc.)?

- Are all escape routes from any occupant position, free from obstructions?

- Where persons are not able to directly evacuate from the building, are there identified places of relative safety from which they can be assisted in their escape.

- Can it be confirmed that all means of escape lead directly outside of the building (other than where there are designated places of relative safety)? Note that no means of escape should lead back into a potential fire affected area.

- Are all escape routes designated by appropriate signage, appropriately illuminated and fitted with emergency lighting?

- Does the evacuation strategy rely on control systems to maintain its efficacy (fire damper and door systems, smoke control systems, pressurization systems etc.)? See the following sections for more considerations relating to such systems.

- What are the special provisions for disabled or mobility impaired persons? Where refuges are used, are they suitable and are the procedures for their use unambiguous in the event of fire? See Annex A for some specific considerations.

In conjunction with the above, it should be ensured that the method used to warn persons is appropriate. For information, the method of evacuation may be one of the following:

Total evacuation: This is where all persons are simultaneously evacuated from the building, possibly automatically initiated following detection of a fire condition. The main consideration here is that means of escape can cater for the maximum numbers of persons to evacuate safely and within the specified evacuation period.

Phased evacuation: This is where persons are evacuated in stages. An evacuation alarm may be given in the designated areas or floors whilst an alert alarm is given in other areas. In this case, the passive and active fire protection arrangement should ensure that those remaining in the building are separated from the fire for the full duration of the waiting period and up until they reach a place of designated safety.

Managed / coded evacuation: This is where trained personnel assist with the evacuation process and may involve the use of a coded alarm message to alert relevant building personnel. This will give time to investigate the incident and make appropriate decisions.

Manual evacuation: This is where evacuation is controlled by building personnel only and may vary depending on the incident. Messages may be given via a PA system. This relies on suitable training of key personnel and detailed, approved, procedures and parameters in order to be effective.

Fire detection systems

In the vast majority of cases, a fire detection and warning system will be the cornerstone of the fire strategy. In the case of complex buildings, the system is unlikely to be a simple configuration. For example, the system may comprise a number of different types of automatic detection principle, which may include:

- Point type fire detection (heat / smoke/ multi-sensor/other)

- Beam smoke detectors

- Flame detectors

- Linear heat fire detectors

- Aspirating smoke detectors

- Video based fire detectors

Consequently, the assessment will need to evaluate the extent and type of system and determine its suitability to the factors and risks seen. National standards for fire detection and alarm systems will give detailed guidance. The following should be considered:

- Is there a defined requirement for the type and extent of fire detection and alarm system? For example, is the system required to meet a category?

- Does the assessor believe that the system as seen complies with the above category in terms of both coverage and system type?

- Are there any special hazards which may not be appropriately monitored by the detection system?

- Are there any field detection devices that are inappropriately located, both in terms of effectiveness and maintainability?

- Does the control and indicating equipment show that the system is working properly?

- What fire control systems are initiated by the operation of the fire detection system? (These may include fire warning systems, fire/smoke door and damper systems, controls for smoke extract and control systems, fire suppression systems, etc.). Is there a clear cause and effect logic sequence and does this comply with the fire strategy?

- Can it be confirmed that the systems are being properly maintained?

- Is there a false alarm policy or process in place? Is the false alarm rate under control? (This could be useful information, as a poor false alarm rate can reduce confidence in the detection system and thus undermine the evacuation strategy).

Fire warning signals

There will always be a need to warn persons of a fire. The manner and method of this warning will be dependent on the requirements of the evacuation strategy and possibly other strategies where manned intervention to suppress a fire is required. More information can be obtained from national standards covering fire warning systems. Methods may incorporate one or more of the following:

- Manual warning using bells or voice.

- Alarm bells connected directly to the fire detection and alarm system.

- Alarm sounders connected directly to the fire detection and alarm system.

- Voice alarm system, possibly linked to the fire detection and alarm system.

- Public address system, which may be linked to the fire detection and alarm system or may be stand alone.

- Visual alarm systems that may be used to supplement sound based systems or may be designated for specific occupancy groups.

- Fire telephone systems used to allow voice communication between parties during a fire incident.

- Personal alert systems that may cause a device to vibrate. This may be appropriate for persons with sight or hearing difficulties or for persons located in remote areas of the building not covered by the building alarm system.

Furthermore, fire warning systems will need to be configured to assist with the evacuation strategy (see above). The following should be considered:

- Is there a defined requirement for the type and extent of fire warning system(s)?

- Does the assessor believe that the systems used are fit for purpose?

- Are areas where contractors may work covered by a form of warning system?

- Do the systems comply with the evacuation strategy?

- Are there any field warning devices that are inappropriately located, both in terms of effectiveness and maintainability?

- Does the control and indicating equipment show that the system is working properly?

- Can it be confirmed that the systems are being properly maintained?

Fire suppression systems

Fire suppression systems are normally installed for property protection or for protection of special processes where they could endanger life. Consequently, they are historically not often a requirement for life safety although their life safety potential (particularly in the case of automatic water sprinkler systems) is increasingly being realised. Furthermore, fire suppression can often be used as part of a performance based design to allow for reduced levels of fire compartmentation or for longer travel distances. In such cases, it is important that the assessor is aware of this prior to undertaking the inspection. Typical fire suppression systems may include:

- Automatic water sprinkler systems

- Automatic water deluge systems

- Water mist/fog systems

- Gaseous extinguishing systems (both total flooding and local application)

- Volume inerting systems

- Blanketing systems such as firefighting foams.

- Powder or chemical based systems

The assessment will need to evaluate the extent and type of system and determine their suitability to the factors and risks seen. There are a number of national and international standards that cover one or more variants of fire suppression. The following should be considered:

- Does the use of fire suppression comply with the relevant requirements of the fire strategy?

- Is there a defined requirement for the type(s) of fire suppression system? For example, are the system(s) required to meet a category (e.g. low hazard, ordinary hazard or high hazard)?

- Does the assessor believe that the system(s), as seen, comply with the above Category in terms of both coverage and system type?

- Are there any special hazards which may not be appropriately protected?

- Are there any field suppression devices (e.g. sprinkler heads, valve sets etc.) that are inappropriately located, both in terms of effectiveness and maintainability?

- Where the systems utilise control and indicating equipment, can it be confirmed that the system is working properly?

- Are pump sets, tanks, extinguishing system cylinders etc., appropriately stored and are themselves protected from fire?

- Where total flooding extinguishing systems are used, has a room integrity test been undertaken? If so, is this still valid?

Fire and smoke control

The control of fire and smoke around the building can be achieved by passive means (such as by fire compartmentation) and active means (operation of fire doors and dampers, control of smoke extract systems, etc.) . It is important the assessment views how the fire and smoke control strategy links with other strategies such as the evacuation strategy. Key considerations are:

- Are there any control measures for external spread of fire via walls or via roofs?

- Are there any control measures for spread of fire between connected buildings?

- Can it be confirmed that fire compartments and separations are complete with the ratings of fire doors and shutters and fire/smoke dampers the same as the fire separations in which they are installed.

- Are fire doors/shutters appropriately certified and fitted with smoke seals (if required) and correct ironmongery as appropriate.

- Have fire doors / shutters remained fit for purpose? Is there any deteriorisation showing and are gaps around the doors still within

27

acceptable limits?

- Is there a requirement for the use of fire damper and door actuation systems? If so, are the systems installed and maintained in accordance with relevant standards.

- Is there a requirement for smoke control and extract systems? If so, are these designed to appropriate national standards and is there evidence that they are properly maintained?

- Where smoke control and extract systems rely on the use of smoke curtains, can it be confirmed that these are fit for purpose?

- Are there any special measures for the protection of escape routes from smoke (such as air pressurization systems)? If so, can it be confirmed that they are in accordance with the relevant standard and are properly maintained?

Firefighting systems and firefighter provisions

Firefighting systems may range from first-aid firefighting appliances, such as portable fire extinguishers, through to firefighting water supplies for use by professional fire fighters. These are some of the more general considerations:

Are there sufficient types and numbers of portable fire extinguishers and fire blankets available around the premises? If so, can it be confirmed that they are in accordance with relevant standards and have been properly serviced?

- Are hose reels provided? If so, can it be confirmed that they are in accordance with relevant standards and have been properly serviced?

- Is the building equipped with fire mains water supplies (wet or dry)? If so, can it be confirmed that they are in accordance with relevant standards and have been properly serviced?

- Can the following be confirmed as satisfactory for fire and emergency service intervention:

  o Vehicular access arrangements.

  o Fire service escort arrangement access to fire equipment and systems.

- o Arrangements for fire service access into the building.

- o Arrangements for fire service access to upper or lower levels.

- o Arrangements for fire service access to special areas.

- Is there a requirement to assess information required for the fire and rescue Authority to meet it's requirement for a fire safety audit? Has this been included within any documentation or discussions?

5   POST ASSESSMENT TASKS

Recording of findings

Organisations and individuals competent to undertake fire risk assessments may have an in-house style that may be totally relevant to the building under assessment. This may be acceptable to all stakeholders. However, it is recommended that the style is agreed at an early stage. For complex buildings, there will be a need to describe the findings rather than rely upon tick box sheets, which are unlikely to convey the key issues, and scoring matrix, which on their own will not be able to describe the nuances of the findings. Consequently, a fire risk assessment for a complex building is likely to be a comprehensive report.

Significant Findings and Corrective Actions

There are likely to be aspects found during the assessment that warrants further consideration and/or action to be taken. These are often described as significant findings. In such cases, the report should identify the corrective actions necessary and recommend a timescale when the work is to be completed.

Complex buildings may produce a large number of significant findings, so it is recommended that they are presented in a style to ensure that they are not hidden or misconstrued. Furthermore, significant findings and the corresponding corrective actions could be graded based upon severity.

Post assessment meeting

As well as presenting the report, the assessor should meet with those concerned, including the person(s) responsible for fire safety of the building(s), so that they are made aware of the issues.

Post assessment actions

Following receipt of the report, there is likely to be a need to control and monitor the corrective actions. The action plan is a vital part of the fire risk assessment process.

In such cases a system should be set up to allow each finding to be separately monitored so that they are not missed. Regular feedback will assist in ensuring that the proper actions are taken.

Ideally, control of corrective actions should form a quality controlled process or part of a process. When completed, corrective actions should be formally signed off. If there is likely to be a delay in undertaking the actions, then relevant persons should be consulted to determine if the delay is acceptable. In such cases, temporary additional fire precautions may be required. This could include manned monitoring of identified areas.

Note that post assessment tasks are part of the flow chart as shown in Annex B.

6 FREQUENCY OF ASSESSMENTS FOR COMPLEX BUILDINGS

Given that fire risk assessments for complex buildings are likely to require much more resource than that required for more simple buildings, it is recommended that a full and thorough assessment is undertaken every three years. It is, however, recommended that the frequency of full assessments is agreed with all stakeholders.

In addition to the full assessment, a follow up more frequent assessment could be undertaken. This could focus on any changes in the building, its occupancy and processes. This could be conducted annually or more frequently if necessary. This should still include a site inspection.

ANNEX A: GUIDANCE FOR THE EVACUATION OF MOBILITY IMPAIRED PERSONS

Provision	Considerations	Comment
Staircases	For mobility impaired persons, the important features of stairway design include the type and rise dimensions of the steps; the avoidance of open risers; hand rail arrangements; landing and stairway dimensions and the provision of tactile and visual markings.	A possible provision for limited mobility impaired persons and other forms of disability.
Wheelchair Lift	Should not be designated for use as a means of escape. Furthermore, they should not be installed within means of escape stairways unless that is the only practical option for providing access for disabled people to upper floors.	Where necessary, it is essential that the stairway width required for means of escape is maintained for the whole route.
Refuges	An option where stairways are the only available vertical exit route from a building. Refuges provide a safe area for mobility-impaired persons on a horizontal level to allow persons to assist with their evacuation in the event of a fire. They should be located on all levels of a building where disabled access is allowed. A refuge needs to be dimensioned to allow for one or more wheelchairs to manoeuvre into the space without undue difficulty. In areas where large numbers of MIPs can be expected, the dimensions may have to be increased accordingly to allow for additional wheelchairs.	Satisfactory examples include protected lobbies, corridoors or stairwells. Alternatively, an area in the open air such as a flat roof, balcony, podium or similar place, which is sufficiently protected (or remote) from any fire risk and provided with its own means of escape, would be acceptable.
Ramps	Any ramp provided should conform to the relevant national guidance.	Although ramps are useful for taking MIP from one level to another for moderate changes in level, the dimensions required for taking persons from one floor to another would make them impractical in many situations.
Evacuation Lifts	An evacuation lift must be sited within a protected enclosure consisting of the lift well itself and a protected lobby at each storey served by the lift. Furthermore, it should be provided with a protected route from the evacuation lift lobby at the final exit level to a final exit. There are also special requirements for methods of operation and for the power supplies.	A firefighting lift (which is provided principally for the use of the fire service in fighting fires) may be used for the evacuation of prior to the arrival of the fire service.

ANNEX B: FIRE RISK ASSESSMENT PROCESS FLOWCHART

	TASK	OUTPUT
1	Pre-assessment meeting to discuss objectives for the assessment(s).	Agreed outline scope.
2	Documentation gathering and inspection.	An understanding of the key features of the existing provisions .
3	Gap analysis.	Determination of current risks that may warrant special assessment.
4	Scope and timetable formulation.	An agreed schedule with timetable.
5	On-site assessment.	A record of findings with images recorded as appropriate.
6	Completion of on-site assessment.	De-brief of client / responsible person subject to report.
7	Preparation of report including details of significant findings & observations.	Issue of report and post-assessment meeting.
8	Formulation of corrective action plan.	Corrective action plan agreed and implemented.
9	Corrective action plan completed and signed off.	Confirmation that fire safety and protection provisions are acceptable.

ANNEX C: SAMPLE RISK ASSESSMENT TEMPLATE FOR COMPLEX BUILDINGS

This is one example of a template that can be used for complex buildings and is taken from the Guide. Chapter 4 should be consulted for additional information. The template prompts a written description for each aspect rather than a simple scoring system or tick-list.

Note that rating system in this case is:

S = Significant Finding, (requiring immediate attention)

O = Observation (i.e not of the highest priority but warrants further investigation and possible remedial action)

P = Pass.

The most critical rating for each aspect should be entered for the whole assessment and details of the location(s) of each issue can be provided in the "Details" section or in supporting documentation.

BUILDING DETAIL:	ASSESSMENT DATE:	ASSESSOR(S):
Purpose and Scope of Assessment:		
Main Risk Characteristics:		
Supporting Documentation:		

ITEM CATEGORY	CHECK CRITERIA	RATING (S, O, P)	DETAILS
Fire Safety Management	Are there details of responsible persons for fire safety?		
	Is there a clearly stated reporting structure for fire safety?		
	Are there specific policies for fire prevention and housekeeping including the control of processes and hot works? Is there any evidence that these are adhered to?		
	Are fire drills undertaken? Are these effective and do they result in a review?		
	Are there procedures for the control of maintenance of fire protection systems? Is there any evidence that these are adhered to?		
	Is there any form of training for fire safety (e.g. warden training, use of portable fire extinguishers, contractors site working)?		

ITEM CATEGORY	CHECK CRITERIA	RATING (S, O, P)	DETAILS
Building Characteristics	Is the evacuation route to the assembly point(s) unimpeded? Check for features of the building or its location that could impede evacuation in the event of a fire.		
	Is there a clear strategy for the internal fire compartments / separations? Does the assessment confirm this arrangement?		
	Are openings within the building such as voids, that could allow routes for uncontrolled fire or smoke spread controlled or protected by structural means or by system monitoring and containment?		
	Are the fire properties of external and internal walls and linings, floors, ceilings and roof coverings stated and does the site assessment confirm these to be accurate?		
	Can it be confirmed that nominated escape routes can be kept clear for the duration of the evacuation and, if applicable, firefighting action?		
	What forms of natural ventilation exist? Are they suitable for the expected type and quantities of smoke envisaged?		
	What is the potential for fire and smoke to spread through unknown or concealed routes? If so, are there controls in place?		

ITEM CATEGORY	CHECK CRITERIA	RATING (S, O, P)	DETAILS
Plant & Services	How susceptible are key services to a fire?		
	How could plant contribute to fire, smoke and toxic gas movement throughout the building? Has this been catered for by passive or active fire protection?		
	How easy will it be to access critical services in the event of a fire? Are routes leading towards these critical services protected from potential risks?		
Building Processes	How susceptible are key processes to a fire? In this case, location and separation from other risk areas will need to be assessed.		
	How could each of the processes contribute to a fire or to smoke and toxic gas movement throughout the building? Has this been catered for by passive or active fire protection?		

ITEM CATEGORY	CHECK CRITERIA	RATING (S, O, P)	DETAILS
Evacuation and Means of Escape	Is there an evacuation strategy? How is it documented?		
	Does the assessor believe that the various occupancy groups are adequately considered by the evacuation arrangements as documented and/ or as noted during the inspection?		
	Are occupants knowledgeable of the building, its layout and possibly fire and other exit routes from the building? Where not, is the management process suitable in the event of fire?		
	Are there language and comprehension of alarm message issues and have these been taken into account?		
	For performance based designs, are the ASET and RSET figures defined? Is the difference in value sufficient based on the findings of the assessment? Furthermore, do the ASET and RSET figures allow for aspects such as pre-movement time and pre-evacuation procedures and coded warning systems?		
	Is pre-movement time a potential issue?		
	How have the occupancy numbers and profiles been established? Are they a true representation of what is seen during the assessment?		

	What assumptions have been made about spread of persons throughout the building? Is there a high possibility of skewing of numbers into particular parts of the building?		
	Are the horizontal and vertical means of escape clearly identified on drawings? Does the layout as seen during the assessment meet with these drawings?		
	Are all escape routes free from obstructions? If there are any ramps or changes of level on the horizontal escape routes, are they suitable?		
	Where persons are not able to directly evacuate from the building, are there identified places of *relative safety* from which they can be assisted in their escape?		
	Are there any unusual aspects of the means of escape that demand additional attention (such as travel distance, dead end conditions, etc.)?		
	Can it be confirmed that all means of escape lead directly outside of the building (other than where there are designated places of relative safety? Note that no means of escape should lead back into a potential fire affected area.		
	Are all escape routes designated by appropriate signage, appropriately illuminated and fitted with emergency lighting?		
	Does the evacuation strategy rely on control systems to maintain its efficacy (fire damper and door systems, smoke control systems, pressurization systems etc.)? If so, do these support the fire strategy?		

	Where refuges are used, are they suitable and are the procedures for their use clear in the event of fire?		
	What are the special provisions for mobility impaired persons? Reference could be made to Annex A of this Guide.		

ITEM CATEGORY	CHECK CRITERIA	RATING (S, O, P)	DETAILS
Fire detection systems	Is there a defined requirement for the type and extent of the fire detection system?		
	Does the Assessor believe that the system as seen complies with the above in terms of both coverage and system type?		
	Are there any special hazards which require specialist types of fire detection system? Are these hazards appropriately monitored?		
	Are field detection devices appropriately located, both in terms of effectiveness and maintainability?		
	Does the control and indicating equipment show that the system is working properly?		
	What are the main fire control systems derived from the fire detection system? Is there a clear cause and effect logic sequence and does this comply with the fire strategy?		
	Can it be confirmed that the systems are being properly maintained?		
	Is there a false alarm policy or process in place? Where false alarm problems have been identified, have appropriate actions been taken to remedy the problem(s)?		

ITEM CATEGORY	CHECK CRITERIA	RATING (S, O, P)	DETAILS
Warning Systems	Is there a defined requirement for the type and extent of fire warning system(s)?		
	Does the Assessor believe that the systems used are fit for purpose?		
	Are areas where contractors may work covered by warning systems? Note that visual alarms may be required in areas of high ambient noise.		
	Do the systems comply with the evacuation strategy?		
	Are all field warning devices that are appropriately located, both in terms of effectiveness and maintainability?		
	Does the control and indicating equipment show that the system is working properly?		
	Can it be confirmed that the systems are being properly maintained?		

ITEM CATEGORY	CHECK CRITERIA	RATING (S, O, P)	DETAILS
Fire Suppression Systems	Does the use of fire suppression comply with the relevant requirements of the fire strategy		
	Is there a defined requirement for the type(s) of fire suppression system? Are the system(s) required to meet a category of a relevant standard?		
	Does the assessor believe that the system(s) as seen comply with the above category in terms of both coverage and system type?		
	Are there any special hazards? Are these appropriately protected?		
	Are all field suppression devices (e.g sprinkler heads, valve sets etc.) appropriately located, both in terms of effectiveness and maintainability?		
	Where the systems utilise control and indicating equipment, can it be confirmed that the system is working properly?		
	Are pump sets, tanks, extinguishing system cylinders etc, appropriately stored and are themselves protected from fire?		
	Where total flooding extinguishing systems are used, has a room integrity test been undertaken? If so, is this still valid?		

ITEM CATEGORY	CHECK CRITERIA	RATING (S, O, P)	DETAILS
Fire and Smoke Control	Is there a strategy for fire and smoke control? Is this still relevant to the building?		
	Are there any control measures for external spread of fire via walls or via roofs?		
	Are there any control measures for spread of fire between connected buildings?		
	Can it be confirmed that fire compartments and separations are complete, with the ratings of fire doors and shutters and fire/smoke dampers the same as the fire separations in which they are installed?		
	Are fire doors/shutters appropriately certified and fitted with smoke seals and correct ironmongery as required.		
	Are there any signs of wear and tear on the fire doors? Could this impact on their integrity? Of special consideration should be gaps around door.		
	Is there a requirement for the use of fire damper and door actuation systems? If so, are the systems installed and maintained in accordance with relevant standards.		
	Is there a requirement for smoke control and extract systems? If so, are these designed to appropriate standards and is there evidence that they are properly maintained?		

	Where smoke control and extract systems rely on the use of smoke curtains or other forms of smoke containment, can it be confirmed that these are fit for purpose?		
	Are there any special measures for the protection of escape routes from smoke (Such as air pressurization systems)? If so, can it be confirmed that they are in accordance with the relevant standard and are properly maintained?		

ITEM CATEGORY	CHECK CRITERIA	RATING (S, O, P)	DETAILS
Firefighting systems and Firefighter assistance	Are there sufficient types and numbers of portable fire extinguishers and fire blankets available? If so, are they in accordance with relevant standards and have been properly serviced? Have persons been trained in their use?		
	Are hosereels provided? If so, can it be confirmed that they are in accordance with relevant standards and have been properly serviced? Have occupants been trained in their use?		
	Is the building equipped with fire risers/droppers (wet or dry)? If so, can it be confirmed that they are in accordance with relevant standards and have been properly serviced?		
	Can the following be confirmed as satisfactory for Fire and emergency service intervention: - Vehicular access. - Fire service escort arrangement and access to fire equipment and systems. - Arrangements for fire service access into the building, to upper or lower levels and to special areas. - Special requirements to assist with risk audits (where deemed to be in the scope).		

ANNEX D: USEFUL REFERENCE DOCUMENTS

For use in the United Kingdom and where no similar national standards exist, the following British Standards may be useful when planning a fire risk assessment for complex buildings. The latest editions should apply:

BS PAS 79: Fire risk assessment – guidance and a recommended methodology
BS PAS 911: Fire strategies – guidance and framework for their formulation
BS 7974: Application of fire safety engineering principles to the design of buildings
BS 8300: Design of buildings and their approaches to meet the needs of disabled people — Code of practice.
BS 9999: Code of practice for fire safety in the design, management and use of buildings
BS ISO 22301: Societal security. Business continuity management systems. Requirements

ABOUT THE AUTHOR

PAUL BRYANT is a British fire strategist, a chartered fire and electrical engineer and a business entrepreneur. He has built up an international reputation for his approach to the subject of fire engineering and for the fire codes and guidance documents he has written. Paul started his career with a London based fire insurance organisation known as "The Fire Offices' Committee".

He then moved to the UK's Loss Prevention Council before taking on the key role of Head of Fire Engineering at London Underground – the oldest underground railway in the world. Paul went on to form Kingfell in the mid-1990s and grew it from a single manned operation to a multi-million pound "fire" one stop shop – combining consultancy and systems engineering. He also explored combining fire engineering with crisis management.

Paul divested his interests in systems engineering in 2011 and now concentrates on developing a specialist international fire and risk consultancy for complex buildings and infrastructures. He continues to lecture on his chosen field around the world.

Kingfell offers a range of specialist fire and risk services. Check them out at www.kingfell.com.

www.ingramcontent.com/pod-product-compliance
Lightning Source LLC
Chambersburg PA
CBHW071645170526
45166CB00003B/1443